Searching For the Smallest Particles

Harcourt
SCHOOL PUBLISHERS

Orlando Austin New York San Diego Toronto London

Visit *The Learning Site!*
www.harcourtschool.com

What Is the Universe Made of?

If you have ever wondered what our universe is made of, you are not alone! Throughout history, people have asked this question. When you look around, you see many kinds of materials. Water, rocks, trees, grass, and even stars are all different. But why? What are they made of?

Over 2000 years ago, a scientist in Greece named Democritus was sure that he had the answer. Everything is made of atoms! Democritus believed that atoms were the smallest particles and could not be broken apart. He called them *atomos,* which is Greek for "uncut."

Democritus had no microscope, and he did not have a chemistry lab. Everything he believed about matter came from his imagination. The atoms that Democritus imagined were much different from the atoms that you have studied. You know that there are many kinds of atoms. Carbon atoms are different from iron atoms. Zinc atoms are different from oxygen atoms. They all have different numbers of protons. Democritus, however, thought that all atoms were the same. The atoms that make up water, he thought, were the same as the atoms that make up rocks. He believed that atoms had different shapes or were put together differently to make different materials. If you broke the materials down, thought Democritus, the atoms would all be alike.

Democritus believed that atoms were the smallest particles of all matter—so small that they were invisible. He also thought that atoms were completely solid, that they had no holes in them, and that one part of the atom was exactly like the other parts.

Democritus was wrong, but for hundreds of years, no one had a more accurate idea about the smallest particles. Without microscopes or an understanding of chemistry, Democritus' ideas made sense. They explained the world that people knew at that time.

Democritus believed that atoms differed from each other only in shape and in the way they were arranged.

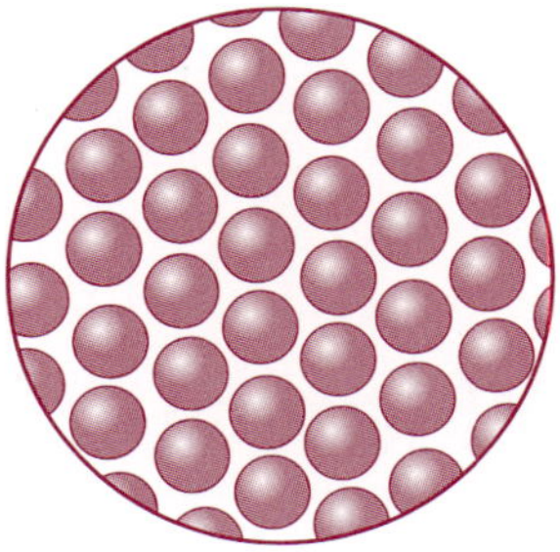

water atoms

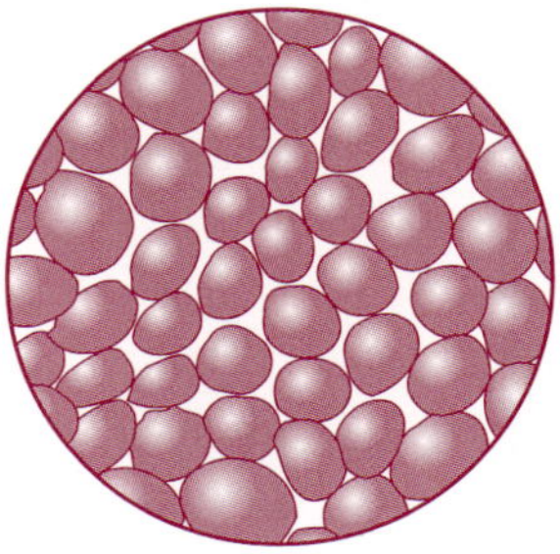

iron atoms

Something Smaller than Atoms?

It may sound strange that people knew so little about atoms. You have heard about atoms many times in your science classes. You may have seen photographs of atoms that were taken using special microscopes. But before the early 1800s, scientists had no way to study atoms. Did atoms really exist? Were they really the smallest particles? People could only wonder about these things.

In 1808, John Dalton made a discovery that changed people's minds. Dalton was a teacher in England. He taught chemistry, but he had never done chemistry research. Instead, Dalton had studied the atmosphere. He showed that air is not just one gas but is made up of many gases. While studying gases, Dalton realized that his findings could be true only if matter were made of atoms. But the atoms that Dalton described were far different from the atoms that Democritus had imagined. Dalton showed that there must be different kinds of atoms.

Dalton's idea was correct—each element has its own type of atom. But Dalton's findings led to another question. If there are different kinds of atoms, what makes them different? Just as people were beginning to accept the idea that atoms were the smallest particles, Dalton showed that they probably were not. There must be something about atoms that makes them different from each other. Could there be even smaller particles?

Electrons

Over time, experiments clearly showed that atoms are made from smaller particles. But what were these particles? How could something so small be investigated?

The particles that make up the atom were discovered through the work of many people. The first of these "subatomic" particles to be found was the electron. By the late 1800s, people knew about electricity, but they did not understand that it involved the movement of charged particles. It was not until 1897 that J. J. Thomson, a physics teacher in England, suggested that electrons were parts of atoms. Thomson did not prove his idea, but his research showed that it might be true. Thomson did not call them electrons. He called these particles *corpuscles*. He thought they were the only subatomic particles. Atoms, he thought, were made up entirely of these corpuscles.

Thomson was not entirely correct, but his ideas were an important step in the search for the smallest particles. It was the first time that someone had shown that atoms may be made of smaller particles.

The Nucleus

There was a problem with Thomson's idea. Electrons have a negative charge, but atoms do not. If atoms were made of electrons, shouldn't all atoms be negatively charged? There must be a positive charge to balance the negative charge. Thomson decided that the electrons must be surrounded by positively charged material. The electrons were like chocolate chips in a ball of cookie dough.

Ernest Rutherford, a scientist who worked with Thomson, designed an experiment to test Thomson's theory. Rutherford decided to shoot a beam of positively charged particles (atoms without electrons) toward a thin sheet of gold foil. If Thomson were right, the beam should pass almost straight through the gold atoms. As he expected, most of the particles did pass through the gold atoms. But to his surprise, some of the particles bounced back. Rutherford repeated the experiment many times. Because of the way the beam bounced back, Rutherford realized that an atom must have a very small, positively charged center, or nucleus. The nucleus, he said, was a ball of protons. The electrons were circling around this nucleus.

Protons, Neutrons, and a Cloud of Electrons

About 20 years after Rutherford discovered the nucleus, however, one of his students noticed something unusual. Like Rutherford, James Chadwick was shooting positively charged particles toward atoms of the gas beryllium. This time, however, he found that particles without a charge were coming from atoms that had been split apart. Chadwick had discovered neutrons, subatomic particles with no charge. The nucleus was not just a positively charged ball of protons. It was made of protons and neutrons.

Over time, ideas of how electrons moved around the nucleus of protons and neutrons changed. At first scientists thought that the electrons moved in set orbits around the nucleus. Electrons are now believed to move in a cloud around the nucleus. By the 1930s, the basic structure of the atom was known. Electrons, protons, and neutrons are the particles from which atoms are made, and scientists believed that they had found the smallest particles that could not be broken apart.

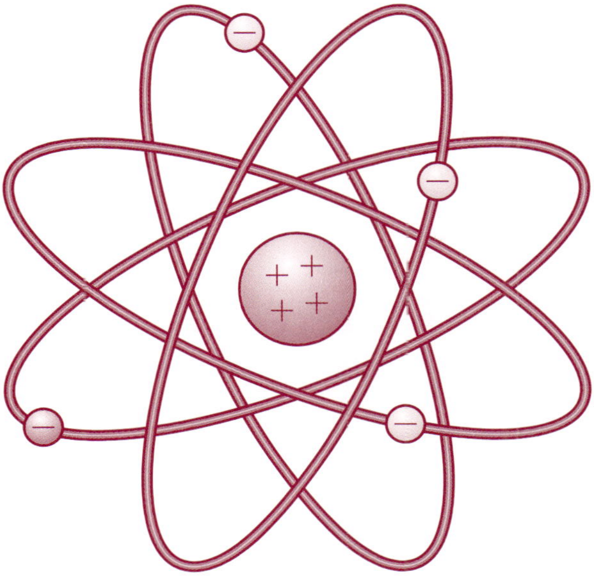

Nuclear model of the atom

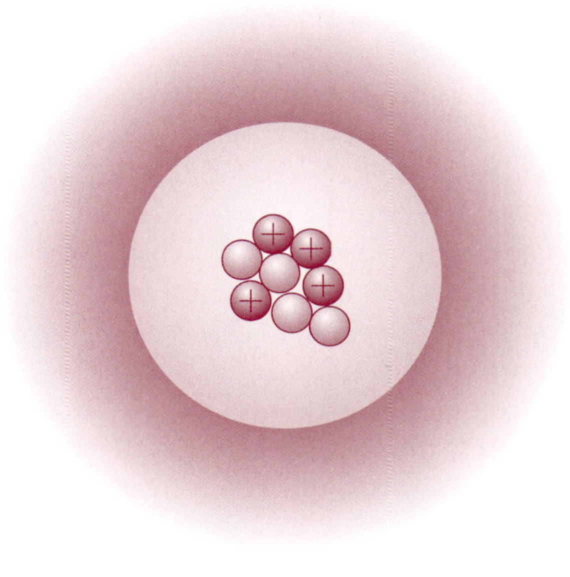

Electron cloud model of the atom

Particles from Space

Every second of every day, high-speed particles from space blast through your body. If you had a detector, you would probably record several of these particles every second. These particles are called cosmic rays. This constant stream of particles will not hurt you, but it is always present. Scientists have known about cosmic rays since the early 1900s, but they understood little about them. Today we know that cosmic rays are mostly high-energy protons along with some electrons and other particles.

In 1932, a scientist in California named Carl Anderson was studying cosmic rays. He was using a device called a cloud chamber to detect the rays. Anderson's chamber was filled with water and alcohol vapor. If he suddenly lowered the pressure inside, cosmic rays passing through would leave a trail of tiny droplets. At the same time that Anderson lowered the pressure, he took a picture that showed the droplets. Anderson also placed a magnet next to the chamber. When a charged particle passed through the chamber, its path would be bent by the magnetic field.

One day Anderson noticed something strange in one of his photographs. One of the particle paths was bent in a different way. He could tell by how much the particle curved that it had the same mass as an electron. But the path showed that it had a positive charge, like a proton. Other experiments showed the same thing. Anderson had discovered another particle! Today this particle is called a positron, or a positively charged electron.

The positron is not one of the particles that make up an atom, but its discovery was important for another reason. It made scientists wonder whether there were more particles that they had not yet discovered. Once again, they began thinking about the smallest particles.

High-Energy Particles

Cosmic ray studies led to other discoveries. Several years after finding the positron, Anderson found another particle in his cloud chamber photographs. This particle had a negative charge, like an electron, but it was much heavier. Today the particle is called a muon. In the late 1940s, researchers discovered a particle that is now called a pion. This particle was much heavier than the electron, but it could have either a positive charge, a negative charge, or no charge at all.

From time to time, scientists make discoveries that start an entirely new way of thinking. Anderson's discovery of the positron and muon prompted new thought. Scientists had thought that the electron, proton, and neutron were the smallest particles. They thought that these were the only particles that make up atoms. Now, however, scientists were beginning to realize that there may be many more subatomic particles. But how could they find more of these particles?

For a long time, the only way to find new particles was through cosmic rays. Why didn't scientists detect the particles in laboratory experiments? The answer was found high on the tops of mountains. Scientists began to wonder whether the atmosphere was blocking some particles from reaching Earth's surface. If so, then more particles should be found at higher altitudes, where the atmosphere was thinner.

Detectors placed on mountaintops showed that these ideas were correct. Not only did scientists detect particles more often, but they also detected "showers" of particles. Particles from space were colliding with atoms in the atmosphere. These collisions were creating new particles, such as muons and positrons, that could not be seen in laboratories at lower altitudes. The particle showers also showed that the energy of the original particles must be tremendous.

One theory about cosmic rays was that particles came from stars that exploded in space. The energy released during these explosions was enough to break apart atoms and send high-energy particles away from the explosions. Some of these particles struck atoms in Earth's atmosphere, causing the particle showers scientists had detected.

Using cosmic rays to find subatomic particles had some drawbacks. On Earth, the number of particles detected each second was small. Finding something interesting was unlikely. Placing detectors on mountaintops helped, but there were still problems. Scientists had no control over when things happened. They had to wait for particle showers to occur. Naturally, they began to wonder whether they could make their own cosmic rays and control the energy and the timing of particle showers.

Particle Accelerators

Could more particles be found using high-speed particles to break apart atoms? This idea was not new. Rutherford had used this technique when he discovered the nucleus, and Chadwick discovered neutrons by shooting particles toward beryllium. The difference now was energy. Cosmic rays had shown the need to use high-energy particles to find the smallest particles of matter.

Since the 1950s, many large machines have been built to produce beams of high-energy particles. These beams are directed toward various targets, such as hydrogen or helium gas. The high-energy particles in the beam smash into the target atoms. The atoms break apart in a shower of subatomic particles. Different types of detectors are used to observe these particle showers.

The machines used to produce the beams of high-energy particles are called accelerators. *Accelerate* means "to change speed." The accelerators increase the energy of particles by increasing their speed. The particles are "pushed" along by electric fields. Only charged particles can be accelerated in this way because only charged particles can react to, or can be "pushed" by, an electric field.

Charged particles also react to magnetic fields. Most accelerators use giant magnets to keep the beams of charged particles moving in a certain direction. In some accelerators, the beam is guided along a straight, or linear, path. These are called linear accelerators. Other accelerators guide the beams in a circle.

Linear Accelerator

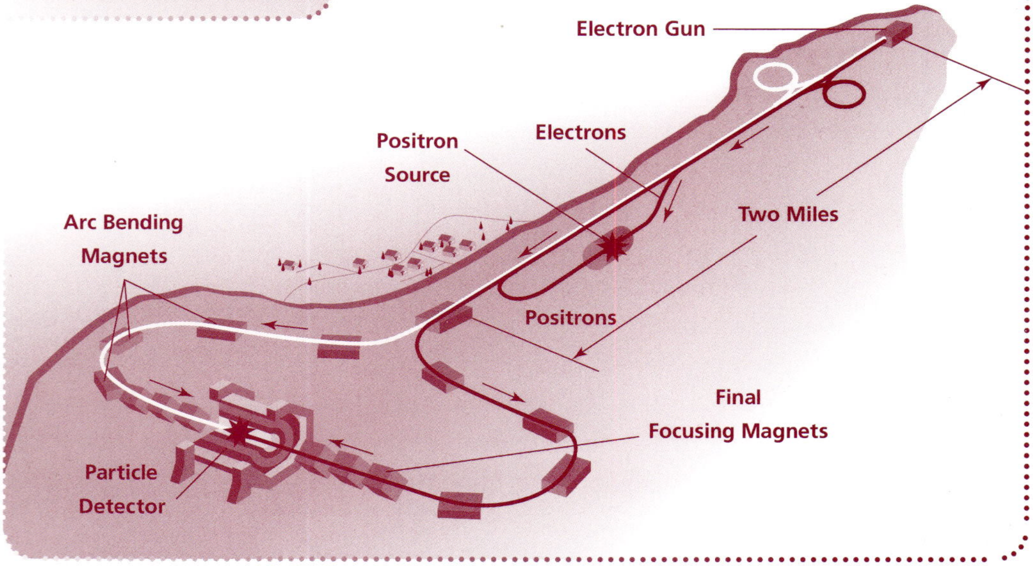

Linear Accelerators

The simplest type of accelerator is linear. An electric field pushes the particles along a straight path. In some cases, the particles strike a target. In other cases, two beams of particles move toward each other and collide. Linear accelerators may also be used as an "injector" to move particle beams into a circular accelerator.

The longest linear accelerator in the world is located at the Stanford University Linear Accelerator Center (SLAC) in California. This accelerator pushes electrons along a 3.2-km (2-mi) path. Electric fields along the path increase the speed of electrons by pulling and pushing on them. SLAC has been used to study protons and neutrons. In recent years, new particles have been discovered at SLAC by colliding a beam of electrons with a beam of positrons.

Circular Accelerators

The energy of the particles that move through linear accelerators depends on the length of the accelerator. To have higher energies, longer accelerators are needed. Circular accelerators solve this problem. The particle beams move through a circular tube many times. Each time they go around the circle, the beams gain more and more energy as they move through a series of electric fields.

Protons are much heavier than electrons. When protons are pushed along a circular path, they lose less energy than electrons. This is one reason that protons are often used in the particle beams of circular accelerators. Like electrons, protons are charged. Their speed can be increased as they move along a series of electric fields. Proton beams can also be guided by magnets placed around the circular path.

The most powerful circular accelerator in the world is the Tevatron, located at Fermilab in Illinois. The Tevatron accelerates protons through a 6.3-km (3.9-mi) circular tube. About 1000 magnets guide the protons along the path. The proton beam circles again and again until it reaches the speed and energy needed for the experiments. The beam is then focused along a straight path that leads to a target.

Another important accelerator is at Brookhaven National Laboratory in Upton, New York. This accelerator has a 3.9-km circular path and uses over 1700 magnets to guide the beam. The particle beam in this accelerator is made of gold atoms that have had their surrounding electron clouds removed. Two beams travel in opposite directions until they are allowed to hit each other head-on.

Circular Accelerator

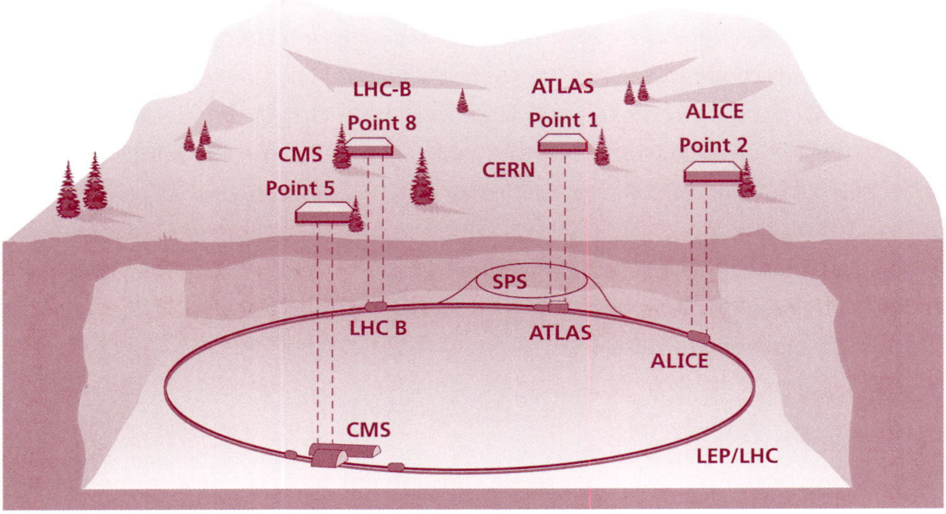

CERN

The search for the smallest particles is very different today than it was when electrons, protons, and neutrons were first discovered. Today, only giant accelerators can produce the high-energy particle beams needed for research. An enormous amount of time and planning goes into each experiment.

The smallest particles can be found only if many scientists work together. Experiments are expensive, and therefore several nations must pool their resources to do the research. The Center for European Research (CERN) is an example of scientists working together to search for the smallest particles. About 6500 scientists from 80 countries work together on the experiments at CERN.

CERN has had many types of accelerators. The newest accelerator at CERN is the Large Hadron Collider (LHC). The LHC has a circumference of almost 27 km (17 mi). It is designed to produce collisions of either beams of protons moving in opposite directions or beams of the nuclei of lead atoms. With particle beam energies higher than any ever used before, the LHC is expected to uncover more details about the smallest particles of matter.

Detectors

When particle beams collide with each other or with a target, machines called detectors can show what happened. To identify a subatomic particle, scientists must know its mass and its charge. They can find out a particle's mass if they know its speed or energy.

Two main types of detectors measure these properties. Tracking detectors look at the path that particles take. If a particle moves through a magnetic field, for example, the curve of its path shows whether its charge is positive or negative. The amount that a particle's path curves over a certain distance gives information about the particle's speed and mass. The cloud chamber that Anderson used when he discovered positrons was a tracking detector. The other main type of detector is a calorimeter. It is used to measure the energy of particles. A material is placed in the path of the particle beam so that it blocks the particle beam. Scientists measure how far particles move into the material before they stop. Particles with greater energy are able to travel farther.

Understanding the smallest particles means that scientists must use both trackers and calorimeters. Even though particle beams are small, the detectors used to look at the collisions include a lot of equipment. The entire detector setup is often the size of a house! Three common detectors used today are wire chambers, Cherenkov detectors, and scintillation detectors.

Wire Chambers

A wire chamber detector is a metal box with wires stretched side-by-side across it. The wires are positively charged. If electrons with a negative charge pass nearby, they are pulled toward the positively charged wires.

The chamber is filled with a gas. When a charged particle moves through the gas, it pulls electrons away from the gas atoms. These electrons are then pulled toward the wires. The wires are attached to devices that can record the number of electrons that hit each wire. They also record which wire was hit. In this way, scientists can determine the paths of particles that were produced during an experiment.

One type of wire chamber is a drift chamber. Instruments measure the time it takes for electrons to move toward the wires. This gives more information about particle paths. Drift chambers are often used because they provide so much information about particle paths.

Wire Chamber

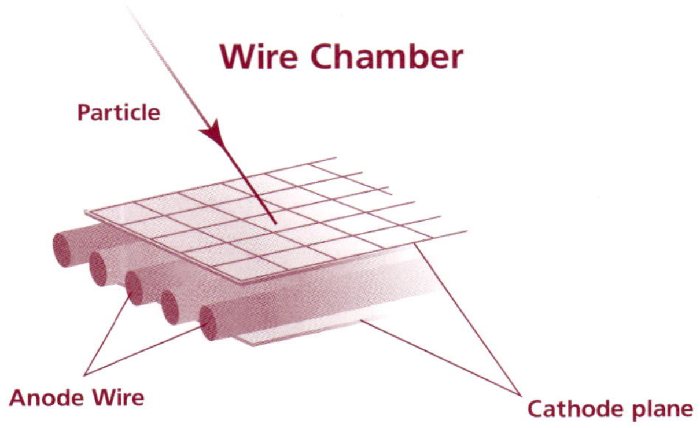

Scintillation Detector

Cherenkov Counter

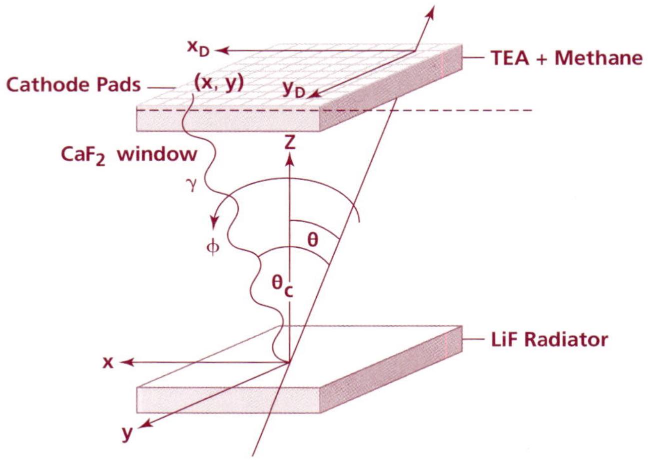

Cherenkov Detectors

You may have heard that nothing travels faster than light, but this is only partly true. In empty space, nothing does travel faster than light. When light travels through matter, however, it slows down slightly. It is possible for some high-energy particles to travel through the matter faster than light does. In 1934, Pavel Cherenkov, a Russian scientist, noticed something unusual when this happened. The particles emitted a faint blue light.

Today, scientists use this light to detect particles. In a Cherenkov detector, a particle passes into a small container. The material (either solid, liquid, or gas) chosen to fill the container depends on the types of particles scientists expect to see in the experiment. Special instruments record the light that is emitted by the particles as they pass through the container.

Cherenkov detectors are useful for recording particles that pass through very quickly. The direction in which the light is emitted is useful in identifying the speed and direction of the particles.

Scintillation Detectors

Light is also used to locate particles in scintillation detectors. Scintillation detectors are made in different ways. Some are slabs of plastic or crystal, and others are containers filled with liquids or gases. Some scintillation detectors are bundles of plastic fibers. Each of these is made of special materials called scintillators.

When high-speed particles pass through a scintillating material, some of the particle's energy is transferred to the scintillator. The energy is changed by the scintillator into short bursts of light. Devices located near the scintillating material collect this light. Information about where the light occurred can help scientists learn about the particle that caused it. The brightness of the light shows how much energy the particle had.

Measuring Light

Detectors often use photomultiplier tubes to record the light. These are used for both Cherenkov detectors and scintillators. The tubes contain a material that is sensitive to light. Light causes atoms in the material to emit electrons. The tubes also contain a series of charged metal plates. When electrons hit the plates, they cause the plates to emit more electrons. These then hit other plates, causing them to emit electrons. The result is a single burst of light that can result in enough electrons to be measured.

Quarks

Experiments with cosmic rays showed that there are more subatomic particles than just the proton, neutron, and electron. Using large accelerators, scientists have been able to discover hundreds of particles! Studies have shown that most of these particles are made of even smaller particles.

Scientists have identified particles called quarks that they believe are some of the smallest particles of matter. As far as scientists know today, quarks cannot be broken apart, and they are not made from other particles. Studies have also shown that quarks cannot exist alone. They always exist with one or more other quarks.

Six types of quarks have been found. They have been given the unusual names of up, down, strange, charm, bottom, and top quarks. These particles are different from one another in their charges and their masses. Unlike the proton and the electron, quarks do not have whole-number charges. The up, charm, and top quark each have a $^{+}\frac{2}{3}$ charge. The down, strange, and bottom quark each have a $^{-}\frac{1}{3}$ charge.

Particles in the nucleus of atoms are made from quarks. A proton is made from two up quarks and one down quark. If you add these, you find that the total charge of these quarks is $^{+}1$. As you know, $^{+}1$ is the overall charge of the proton. A neutron is made from two down quarks and one up quark. The total charge of these quarks is 0. This is the overall charge of the neutron.

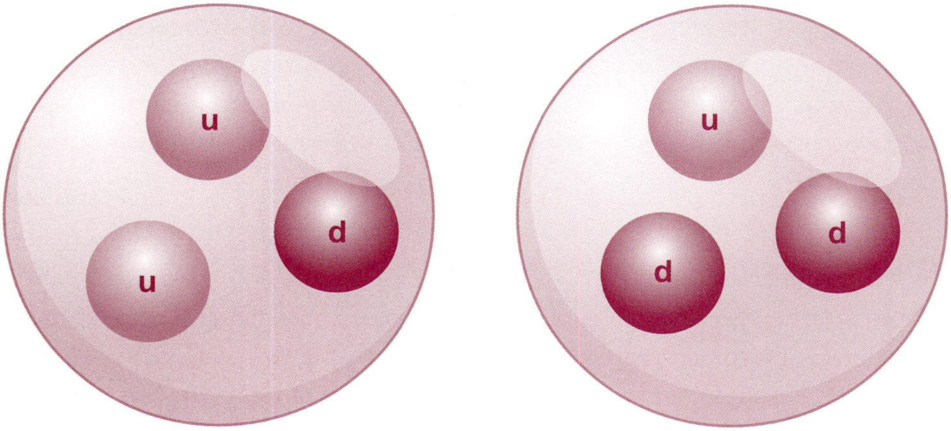

Proton **Neutron**

Leptons

Experiments have also shown another type of particle that is now believed to be one of the smallest possible particles. Leptons have whole-number charges. J. J. Thomson proposed in 1897 that the electron was one of the smallest particles of matter. Today, experiments with huge accelerators have shown that he was correct. The electron is a lepton. As far as scientists know, it cannot be broken apart.

As with quarks, six types of leptons have been identified. They are the electron, the muon, the tau, the electron neutrino, the muon neutrino, and the tau neutrino. Neutrinos have a charge of zero. The other leptons have a charge of –1.

Antiparticles

Chadwick's discovery of the positron, a positively charged electron, was the first known example of an antiparticle. A particle and its antiparticle are identical except that they have opposite charges. If they collide, both are destroyed, creating a burst of light.

High-energy experiments have found an antimatter partner for each of the quarks and leptons. Matter made from antiparticles is not known to exist naturally, but it can be created in a laboratory.

Force Carriers

If you toss a ball into the air, what causes the ball to fall? You know that the answer is the force of gravity. But how can a force act on an object without touching it? Think about what happens when you place two magnets near each other. The magnets seem to push away or pull toward each other, even if they do not touch. Gravity and magnetism are examples of forces that act at a distance. Objects do not have to touch to experience the effects of these forces.

Scientists began using large accelerators to search for particles of matter that could not be broken apart. They found something they did not expect—there are other particles that are not part of matter. These particles are responsible for carrying forces between particles of matter.

Four types of force carriers are believed to exist. A photon carries electric and magnetic forces. Gluons are the carriers for the strong nuclear force. This force holds quarks together in the nucleus. Particles called the W and the Z are carriers for the weak nuclear force, another type of force in the nucleus. Scientists believe that a particle called the graviton is a force carrier for gravity, but this particle has not yet been observed.

The Smallest Particles?

Scientists now believe that they have identified the smallest particles. The quarks and leptons combine to form all other matter in the universe. The force carriers hold the particles together. All of the things that happen in the universe are possible because of force carriers moving among quarks and leptons.

But scientists still wonder whether these are really the smallest particles. Democritus thought that atoms were the smallest particles. In the early 1900s, the electron, proton, and neutron were believed to be the smallest particles. Using high-energy particle beams to break apart atoms uncovered smaller particles. If particle beams with higher energies are used, will they show that quarks and leptons are made up of even smaller particles? This is a question that cannot yet be answered, but scientists will continue searching.

The smallest known particles

quarks	up +2/3	down -1/3	anti-up -2/3	anti-down +1/3
	strange -1/3	charm +2/3	anti-strange +1/3	anti-charm -2/3
	bottom -1/3	top +2/3	anti-bottom +1/3	anti-top -2/3
leptons	electron -1	electron neutrino 0	positron +1	anti-electron neutrino 0
	muon -1	muon neutrino 0	anti-muon +1	anti-muon neutrino 0
	tau -1	tau neutrino 0	anti-tau +1	anti-tau neutrino 0
force carriers	photon	gluon	W+ W- Z	graviton
	electricity magnetism	strong force	weak force	gravity